**NATIONAL GEOGRAPHIC**

School Publishing

# It's Electrifying

## PIONEER EDITION

By Sara Cohen Christopherson

## CONTENTS

# It's Electrifying

Flash! Somewhere, right now, lightning is hitting Earth.
About 40 bolts can strike our planet every second.

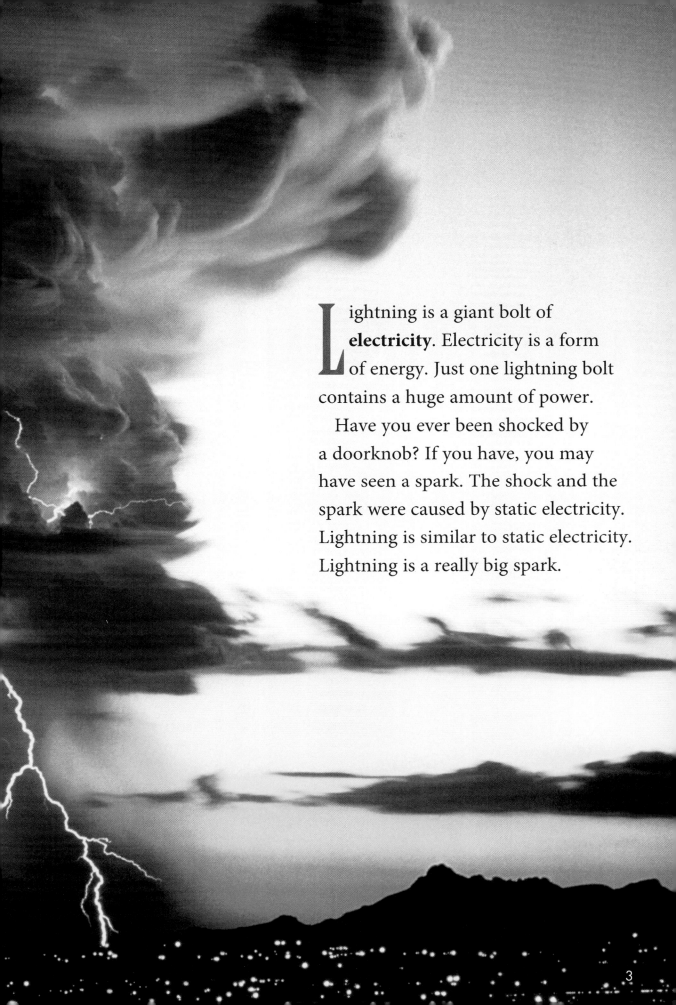

Lightning is a giant bolt of **electricity**. Electricity is a form of energy. Just one lightning bolt contains a huge amount of power.

Have you ever been shocked by a doorknob? If you have, you may have seen a spark. The shock and the spark were caused by static electricity. Lightning is similar to static electricity. Lightning is a really big spark.

**W**here does this electricity come from? This electricity builds up inside clouds. Clouds are made of dust and water droplets. Wind blows the dust and droplets around inside the cloud. This makes a positive charge at the top of the cloud. It also makes a negative charge at the bottom.

The ground below a thundercloud has a positive charge. Lightning flashes between areas with positive and negative charges.

**❶** The top of a cloud has a positive electrical charge.

**❷** The bottom of a cloud has a negative electrical charge.

**❸** The ground below a thundercloud has a positive charge.

**❹** When the positive and negative electrical charges get strong enough, lightning flashes between the cloud and the ground.

**❺** Lightning also flashes between the top and bottom of a cloud.

# Kinds of Lightning

Lightning comes in many different forms. Here are a few kinds you might see in your neighborhood.

**Forked lightning** looks like tree branches.

**Sheet lightning** is a flash of lightning inside a cloud.

**Heat lightning** is a flash of lightning you see from far away.

## A Big Bolt

How big is a lightning bolt? A real bolt of lightning is only 1–2 inches wide. Some lightning is very long. It can be over 100 miles in length.

Lightning is very hot. It is even hotter than the sun!

## Sparking Safety

Lightning is dangerous. It can cause fires. It can also hurt people. Each year, lightning kills about 70 people in the United States. It injures around 300 others.

## Lightning Strikes!

Lightning strikes some parts of Earth more often than other parts. Florida has more lightning than anywhere else in the United States.

**City Lights.** Lightning strikes the Empire State Building in New York City. Its lightning rod protects it.

**Spark of Inspiration.** Benjamin Franklin flew a kite during a thunderstorm in 1752. That helped him prove that lightning is made of electricity.

Over time, people have looked for ways to stay safe from lightning. In the 1750s, Benjamin Franklin came up with an idea. He made a lightning rod. A lightning rod is a piece of metal. It goes on top of a building. A wire connects the metal to the ground. Lightning can strike the metal rod. Then, the electricity travels through the wire to the ground. The building stays safe.

It is not safe to be outside during a thunderstorm. You should go indoors. If there are no buildings nearby, you could go inside a car.

## Watch Out!

- **Check the weather.** Before you go outside to exercise or play, find out what the weather will be like. Stay home if a bad storm is on its way.

- **Don't fool around.** Lightning is powerful stuff. Don't wait until a storm is on top of you. Go inside at the first sign of thunder or lightning.

- **Find shelter.** Porches and open shelters aren't safe during a storm. Go inside a building. If there are no buildings, a car will also do.

- **Stay away from trees.** Standing under a tree might help you stay dry—but it's the last place you want to be in a lightning storm.

# What Makes Thunder Rumble?

Flash! You see a bolt of lightning. Boom! You hear thunder. Why does thunder follow lightning?

Lightning is superhot. A bolt heats the air to more than $23,871^\circ$C $(43,000^\circ$F). Air is made of tiny parts called molecules. Lightning makes these tiny parts of air move quickly apart.

After lightning strikes, the air cools. The tiny parts of air move closer together again. The air moves so fast that it makes a sound. We call that sound thunder.

## An Electrifying Idea

Lightning contains electrical energy. Some inventors have tried to capture the electricity in lightning. They want to use that energy.

So far, no one has been able to capture lightning's energy. Still, some inventors keep trying. Maybe it will happen in the future.

## Shocking, But True

Why is it so hard to capture lightning? One problem is that no one knows where lightning will strike. Another problem is that lightning can be very dangerous.

There are other problems, too. Lightning strikes very quickly, and it has a lot of energy. That makes it hard to store. Also, the electrical energy in lightning changes into other forms of energy. Some changes into light energy. That is the flash of light that you see. Some changes into heat energy. Then heat energy changes into sound energy. That is what you hear as thunder. Once the energy has changed forms, it is even harder to capture.

## Electrical Power

Homes, schools, and businesses use a lot of electricity. For now, nearly all of that electricity is made at a power plant. A power plant has a big **generator**. A generator uses movement to produce electrical energy.

A generator works by turning a copper wire through the north and south poles of a magnet. The movement produces an electric current. The current turns the wire. This produces electricity.

The next time you see a lightning flash, think about what type of lightning you are seeing. And be careful around this form of energy.

**North pole of magnet**

**These magnets produce a magnetic field.**

**South pole of magnet**

**The wire coil spins past the north and south pole of the magnet. This produces electricity.**

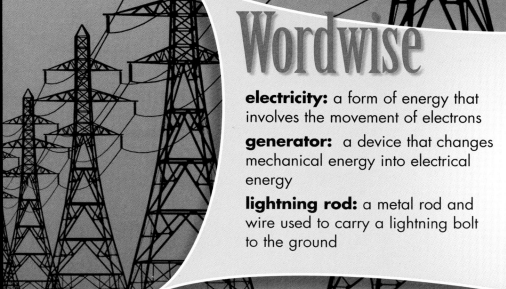

# Wordwise

**electricity:** a form of energy that involves the movement of electrons

**generator:** a device that changes mechanical energy into electrical energy

**lightning rod:** a metal rod and wire used to carry a lightning bolt to the ground

# People Power

Electricity can also be people powered. A person can turn the copper wire in a small generator. This can be useful during a power outage. It can also be useful when a person is far away from an electrical outlet. If you have people power, you don't need batteries, either!

**Light the Night.** These bicycle lights have no batteries. The pedaling powers the lights.

What if you are in a place with no electrical outlets? With people power you can still listen to the radio or make phone calls. You could even use a computer!

Just wind up your mobile phone and you're ready to call.

This radio is powered by a hand crank. Spinning the crank gets the generator going.

Power outage? No problem. Just shake or squeeze one of these.

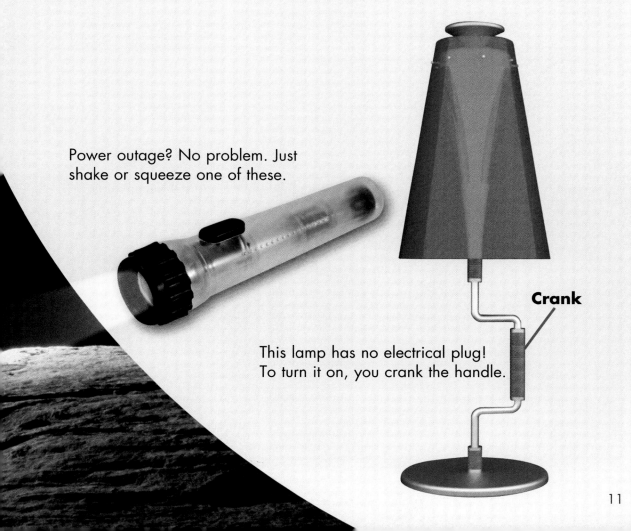

**Crank**

This lamp has no electrical plug! To turn it on, you crank the handle.

# Lightning

**Answer these shocking questions to find out what you learned.**

**1** What is lightning?

**2** What causes lightning?

**3** Why is lightning dangerous?

**4** What is a generator?

**5** How can people power produce electricity?